The Handbook of Office Automation

The Handbook of Office Automation

Ralph Tomas Reilly, Ph.D

iUniverse, Inc.
New York   Lincoln   Shanghai

The Handbook of Office Automation

iUniverse, Inc.

For information address:
iUniverse, Inc.
2021 Pine Lake Road, Suite 100
Lincoln, NE 68512
www.iuniverse.com

ISBN: 0-595-30690-X

Printed in the United States of America

For Anastasia, Aida and Alexander, the A team

Contents

Chapter I

Office Automation

To the novice the subject of automation and its impending ramifications seem almost frightening. At one time perhaps, ten years ago all software applications were cumbersome and not at all user friendly (difficult to understand). Almost all applications had to be constructed around the particular needs of the company. Now however the packages have become more generic and by just selective shopping and conversations with vendors, anyone can pick a suitable software package for their company's needs. Like the old saying "if it already works why fix it." Well there are proven software packages on the market that just need to have installation instructions followed and you're in business. Everything from personal checkbook balancing programs to major accounting packages. Most are designed for ease of installation and can be done without the need of technical personnel. Test allowances are available to permit testing by entering data prior to arrival of the on order system. This testing can be done at Customer Centers or large corporate software branch offices.

Many applications are available through software and service organizations. Public domain is another form of obtaining necessary software. This is available through what are termed bulletin boards, which is simply a listing of software products accessible by modem connections displayed by other users.

With this as the basic introduction to office automation this thesis will now delve into a systematic break' down of the actual steps involved in the developmental processes.

Developmental Program

> ➢ Administration
> ➢ Documentation
> ➢ Installation
> ➢ Product Usability
> ➢ Project Management
> ➢ Information Center
> ➢ Training

ADMINISTRATION

The writing, approval and distribution of organizational standards and procedures to aid in the central control of personal computer systems will depend entirely upon the local environment. In some organizations, standards and procedures are well accepted and readily enforced. In organizations at the other end of the spectrum, corporate designated procedures carry little weight. Even in the latter cases, however, well written procedures that do not demand too much extra effort can be of great-educational value, and can give local management easy directions for preferred actions. They may not be co-operative but they can profitably use the information.

At a minimum, there should be some standardization attempted to two or three hardware vendors, as few types of operating systems as possible, and useful application packages. To allow further proliferation simply increases the cost of operation for everyone, and reduces the vendor support that should be expected.

Similarly, although knowledge about computers is increasing there will always be too little understanding of the best selection reasons and methods. Procedures are needed to help people define their requirements and select appropriate equipment and systems. Rule of thumb for selections and standard methods and procedures are invaluable for training people and for giving manager critical information rapidly and accurately.

A User Support Plan should be carefully analyzed, formally approved and widely understood. It can be the key to gaining acceptance for the approved procedures for managing computer systems and thus helping to optimize the investment in them.

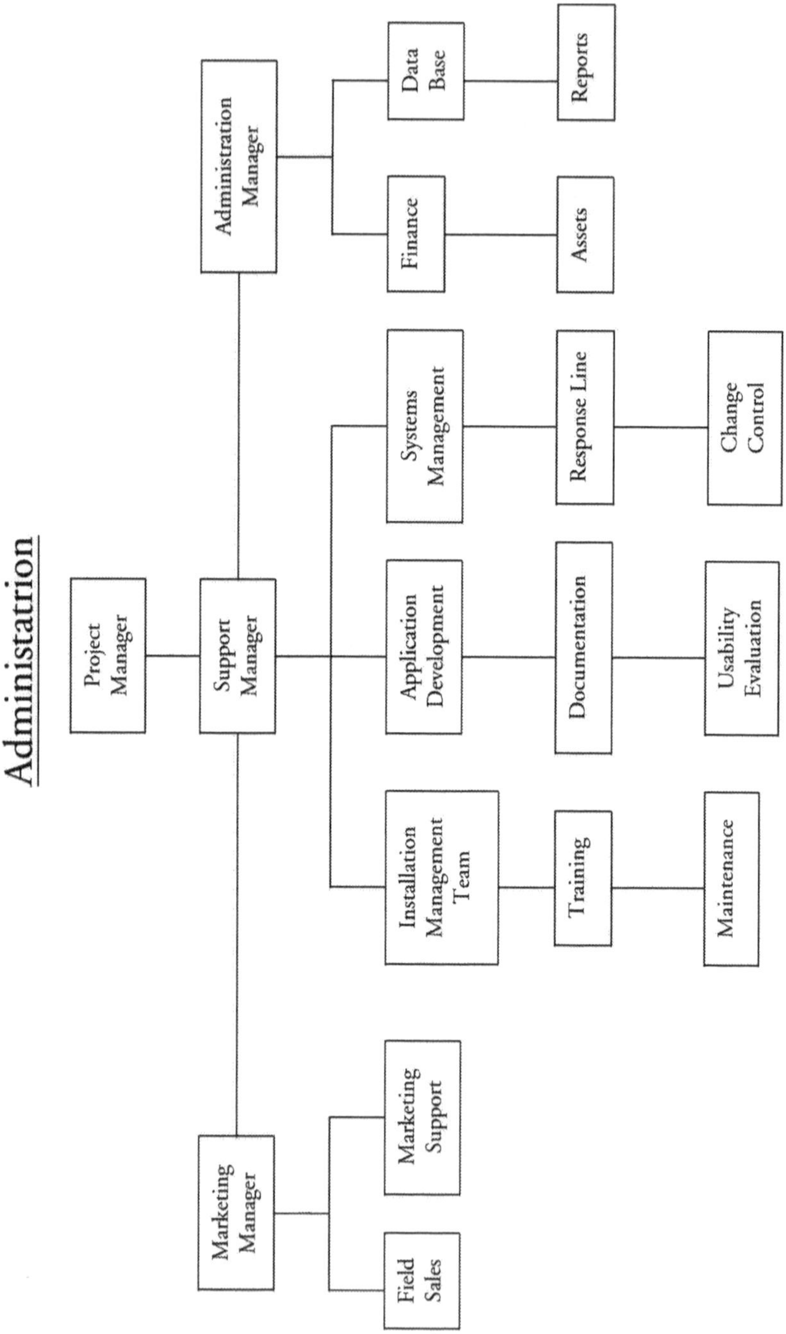

Administratrion

```
                                    Administration
                                      Manager
                                         |
                          +--------------+--------------+
                          |                             |
                       Finance                       Data
                                                      Base
                          |                             |
                       Assets                        Reports


Project          Support                    Systems
Manager          Manager                    Management
    |               |                           |
    +---------------+                      Response Line
                    |                           |
        +-----------+-----------+           Change
        |           |           |           Control
  Installation  Application  
  Management    Development  
     Team          |         
        |      Documentation 
    Training        |        
        |       Usability    
  Maintenance   Evaluation   


Marketing
Manager
    |
    +---------+
    |         |
 Field     Marketing
 Sales     Support
```

FEASIBILITY ANALYSIS

An analysis to estimate the cost and benefits of a proposal and the expectation of a successful project before funds are planned for a project. If a substantial effort is to be put into the development and use of computer systems, an early effort should be put into an analysis of the probable factors involved, usually, before the planning cycle is complete.

Objectives are to determine where the available state of the art matches the probable requirements. To determine whether it is feasible to expect that the objectives can be met within the probable investment and personnel restraints.

The feasibility analysis should address:

1. The principal applications that are being considered.
2. The availability of application systems or software packages.
3. The level of effort for installing and operating the systems.
4. The probable size and configuration of the equipment.
5. The probable costs of system, equipment and operation.
6. The perceived advantages of the system, including financial benefits.
7. The relationship to existing systems and to the central Information Services organization.
8. Levels of approval that will be needed.

PLANNING

Planning for computer systems concerns the application of computer technology to all parts of the operation where both free-standing and network connected computers are used for increasing productivity. This will include analysis, technical calculations, office automation, process control, etc.

The objective is to ensure the appropriate integration of our present and future system and communications.

The planning process should address:

1. Cost and benefit analyses.
2. Impact assessments on the present facilities, systemsand networks.
3. Integration with existing information systems and networks.
4. Coordination with existing information systems and networks.
5. Use of compatible languages and operating systems.
6. Use of existing or purchasable software packages with possible modifications.
7. Organizational and training issues.

The planning process should be handled individually in operating units, and then submitted to the Information Center Group for comment.

EQUIPMENT ADMINISTRATION

The equipment administration plan is an Information Center's method of controlling the delivery and installation of computer systems. The Information Center will provide checklists to coordinate site evaluation, equipment delivery, installation and inventory control.

IMPLEMENTATION PROJECT

Through the use of checklists produced by the information center and related documentation supplied by the vendor, the Information Center will control physical planning, future related equipment updates of both hardware and software. The tracking and shipping of equipment the related equipment will also be accounted for.

SITE EVALUATION

The Information Center will provide the implementation sites with a Document of Instruction, which will arrive at the implementation site before the computer is delivered. This Document of Instruction will detail the physical site requirements for the computer and or workstations.

A) DETERMINE HARDWARE LOCATION

Key management or the technical coordinator will determine the physical site location for the computer and/or workstations based on the physical requirements expressed in the Document of Instruction.

B) SITE PREPARATION CHECKLIST

This document details the specifications and conditions that must be met for installation of the computer system and related workstations, printers and other devices.

C) TIME FRAME FOR SITE COMPLETION

A person at the site will notify the Information Center of the projected date for site completion. If the particular date previously agreed upon cannot be reached then that person will notify the Information Center to make changes to the equipment schedule.

EQUIPMENT DELIVERY

The ordering of the computer and computer related peripherals are the Information Center's responsibility. All equipment and supplies will be ordered after the completion of site evaluation.

DELIVERY CONTROL

This is the Information Center's method of tracking the progress of all equipment and supply orders.

A) BACKLOG: The Information Center will verify with the vendor on a regular basis the status of each order.

B) SHIPMENT CONTROL: The Information Center will determine the shipping status for the computer and computer related equipment and notify the receiving sites of shipping and delivery dates.

C) A person on site will perform required validations and inform the Information Center upon receipt of order.

INVENTORY CONTROL

The Information Center will maintain and update an inventory list of all computer related supplies and equipment. This inventory list will be completed by site personnel and a copy will be sent to the Information Center. This will include all hardware, software with serial numbers received at the site.

Site Preparation Checklist

_____
Overall Completion Target Date

_____
Customer # and Name

_____
Schedule Ship Date

Equipment on Order:

Model _____

Model _____

Model _____

	Target Date	Date Completion
	_____	_____

Cables Installed_____

Electrical Wiring Completed_____

Physical Layout Completed_____

Communications Lines Installed_____

Furniture Ordered_____

Shipped and uninstalled Report

_____

Report Date

Order #	Customer # and Name	Date Shipped	Schedule Install Date	Days Since Ship 15 20 30	Actual Install Date	Comments
_____	_____	_____	_____	_____	_____	_____

Sort By:

Date Shipped

Scheduled Install Date

Actual install Date

Days Since Shipment (High to Low)

On Order Report

Report Date

Order #	Model #	Date Ordered	Requested Ship Date	Scheduled Ship Date	Customer # & Name	Comments
___	___	___	___	___	___	___

Sort by:
      Order #
      Requested Ship Date
      Scheduled Ship Date
      Customer Number and Name

ADMINISTRATIVE BACKLOG CONTROL CHECKLIST

o <u>Pre Sale Procedures</u>
   — Have sales kits contents been decided upon for sales representatives (to include contracts, order forms, documents of under-standing, etc.)?
   — Have contracts been written, received by a legal department, and printed?
   — Are order rates being compared to forecast?
o Have procedures for returning orders been reviewed with Sales to assure timely and accurate return of orders?
o Are order confirmation meetings or contacts with customers being done (allow sufficient time to coordinate changes, etc. to IBM— 120–90 day)?
o Are penalties to be charged for late alterations or cancellations?
o Is there a procedure in place to assure coordination of software and additional equipment, such as modems, printers, etc.
o Have procedures been established to communicate to the customer the specific requirements for site preparation and has a control been put in place to assure that the site has been prepared before equipment is delivered?
o Has necessary training (customer and installer) been completed?
o Have delivery procedures been established (to include "ship to" locations and addresses)?
o Is the customer aware of his responsibilities regarding insurance and other liabilities?
o Has customer placed the equipment on maintenance?
o Has the installation been reported to the proper administrative group to assure inventory update

o <u>Billing and Accounts Receivable</u>
   — Have invoicing and posting procedures been established?
   — Is the procedure to get "installed" information to the billing department adequate?
   — Have "credit checks" been done?

— If a "third party" is used for financing, have procedures been established to communicate appropriate data (customer information, machine serials, dates, etc).

o Inventory Control
  — Has an administrative database been established to track and control the installed inventory?
  — Has a procedure been developed to effectively communicate software and hardware changes to the customer?

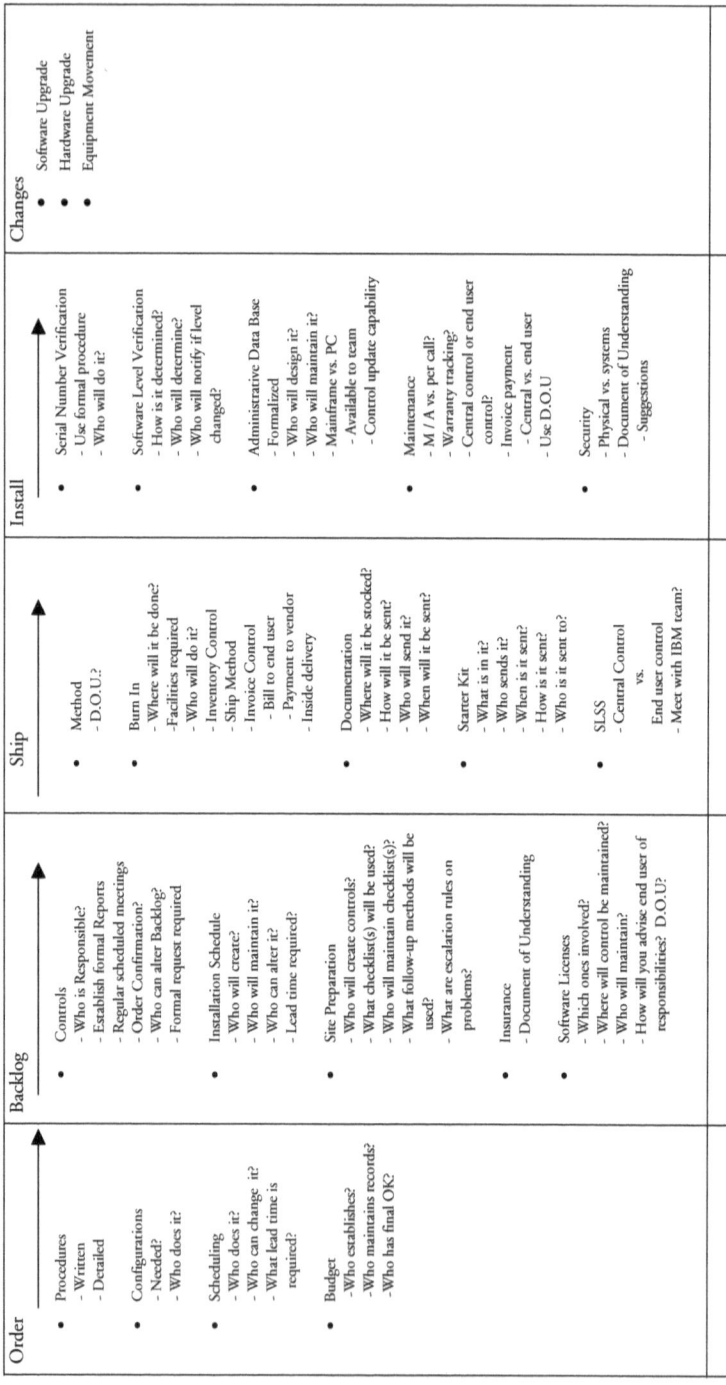

Order	Backlog	Ship	Install	Changes

• Procedures
  - Written
  - Detailed
	• Controls
	  - Who is Responsible?
	  - Establish formal Reports
	  - Regular scheduled meetings
		• Method
		  - D.O.U?
			• Serial Number Verification
			  - Use formal procedure
			  - Who will do it?
				• Software Upgrade
				• Hardware Upgrade
				• Equipment Movement

• Configurations
  - Needed?
  - Who does it?
	• Order Confirmation?
	  - Who can alter Backlog?
	  - Formal request required
		• Burn In
		  - Where will it be done?
		  - Facilities required
		  - Who will do it?
		  - Inventory Control
			• Software Level Verification
			  - How is it determined?
			  - Who will determine?
			  - Who will notify if level changed?	

• Scheduling
  - Who does it?
  - Who can change it?
  - What lead time is required?
	• Installation Schedule
	  - Who will create?
	  - Who will maintain it?
	  - Who can alter it?
	  - Lead time required?
		• Invoice Control
		  - Ship Method
		  - Bill to end user
		  - Payment to vendor
		  - Inside delivery
			• Administrative Data Base
			  - Formalized
			  - Who will design it?
			  - Who will maintain it?
			  - Mainframe vs. PC
			  - Available to team
			  - Control update capability	

• Budget
  - Who establishes?
  - Who maintains records?
  - Who has final OK?
	• Site Preparation
	  - Who will create controls?
	  - What checklist(s) will be used?
	  - Who will maintain checklist(s)?
	  - What follow-up methods will be used?
	  - What are escalation rules on problems?
		• Documentation
		  - Where will it be stocked?
		  - How will it be sent?
		  - Who will send it?
		  - When will it be sent?
			• Maintenance
			  - M / A vs. per call?
			  - Warranty tracking?
			  - Central control or end user control?	

	• Insurance
	  - Document of Understanding
		• Starter Kit
		  - What is in it?
		  - Who sends it?
		  - When is it sent?
		  - How is it sent?
		  - Who is it sent to?
			• Invoice payment
			  - Central vs. end user
			  - Use D.O.U	

	• Software Licenses
	  - Which ones involved?
	  - Where will control be maintained?
	  - Who will maintain?
	  - How will you advise end user of responsibilities? D.O.U?
		• SLSS
		  - Central Control
		    vs.
		  - End user control
		  - Meet with IBM team?
			• Security
			  - Physical vs. systems
			  - Document of Understanding
			  - Suggestions	

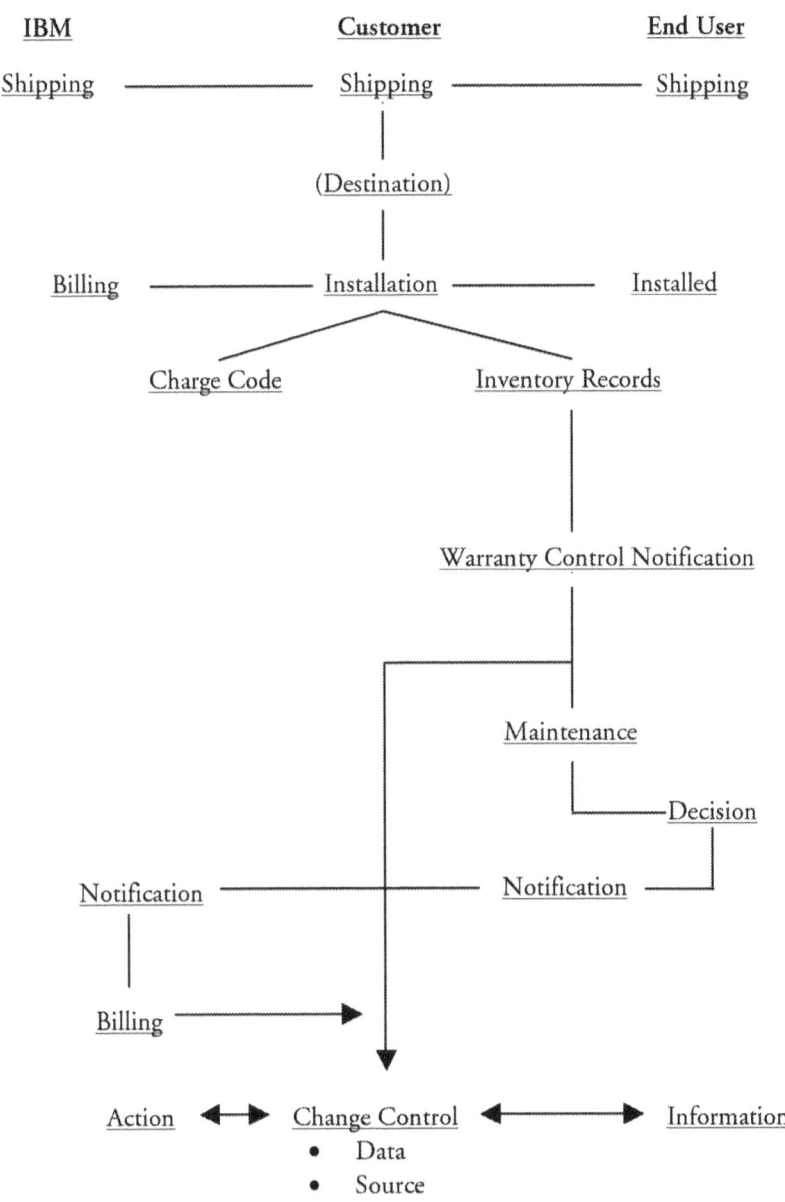

IMPLEMENTATION PLAN

Preliminary Implementation

Project Orientation:
1. Meet with key management
2. Meet with office staff

Site Evaluation:
1. Determine physical site location.
2. Identify and inventory supplies that must be acquired to support the operation environment.
3. Prepare checklists of all items to be considered at the location.
4. Summarize requirements based on recommendations from a contractor.
5. Finalize order volumes and required delivery dates.
6. Determine preliminary training needs.

Contractor Follow Up:
1. Sign contracts with vendors for hardware and software selected earlier.
2. Insure/monitor physical site preparation is complete.

Install Hardware:
1. Follow up on delivery dates with vendor.
2. Organize receipt of hardware.
3. Monitor installation and acceptance testing.

Install Operating System/Other Software
1. Follow up on delivery dates
2. Install software according to instructions.
3. Install the test system
4. Test the software

Install Applications Software:
1. Load application software.

Develop Procedures:
1. Review specific user requirements.
2. Review current procedures.
3. Determine who will perform each procedure.
4. Draft detailed procedures.
5. Review interfaces with existing systems.

Plan Training Program for Users:
1. Confirm user training requirements.
2. Confirm on-going training.
3. Develop employee training schedule.

Train Personnel:
1. Meet with management.
2. Meet with office staff.

Complete Conversion Plan
1. Finalize conversion dates.
2. Document data conversion procedures.
3. Document the overall data conversion controls.
4. Determine how data will be backed up.
5. Review conversion plan with management.

Develop Conversion Procedures:
1. Develop manual conversion procedures documentation.
2. Review developed procedures and refine them.
3. Publish conversion procedures.

Purify Conversion Files:
1. Collect data.
2. Transcribe data.
3. Review results and validate data.
4. Execute procedures and programs to create data files.
5. Review accuracy and completeness of converted data.
6. Maintain conversion files.
7. Maintain the back-up copies of all data.

Project Status:
1. Meet with management.
2. Meet with office staff.

Perform Integration Test:
1. Monitor system performance
2. Verify results.

User Validation:
1. Review implementation plans.
2. Obtain written verification from user and operations
3. Verify all prerequisites are completed.
4. Resolve and open issues.

Implementation

Orientation/Implementation
1. Meet with management on project status.
2. Meet with office staff on project status.
3. Initiate Implementation checklist.

Monitor Production:
1. Provide support during initial production period.
2. Monitor the manual processes.
3. Monitor the automated processes.
4. Record and control variances from design.

Transfer to Operations:
1. Verify that new system documentation is ready for operation.
2. Verify operational personnel are ready to operate the new system.
3. Remove the old system if appropriate.
4. Establish user liaison.
5. Complete maintenance manual.
6. Transfer responsibility to support group.

Post Implementation

Orientation:
1. Meet with management to critique project and determine next steps.
2. Meet with office staff to critique project.

Monitor and Tune System:
1. Review initial design documentation.
2. Review actual system operations data.
3. Identify significant variances.
4. Document and analyze results.

Implementation Charts

Project Schedule

° Preliminary

° Implementation

° Post

Reviewed By:

Review Date:

Project Schedule:

Project Name:

Date Prepared:

Prepared By:

Project Segment / Task	No. Of Persons	Man Days in 2 Day Increments - Not Including Weekends, Holidays, or Vacations (* - Half Days) (** - Full Days)													
Implementation															
Monitor Production	1	**	**	**	**	**	**	*							
Transfer to Operations	1			*	**		*								
Project Critique							*								
Post Implementation															
Orientation								**							
Monitor and Tune System	1							**							
Final Project Status								*							
Total Man Days		2	2	1	1	1	1	1	1					1	

Reviewed By:
Review Date:

Project Schedule:
Project Name:

Date Prepared:
Prepared By:

Project Segment / Task	No. Of Persons	Man Days in 2 Day Increments - Not Including Weekends, Holidays, or Vacations (* - Half Days) (** - Full Days)
Preliminary Investigation		
Project Orientation	-	*
Site Evaluation	1	**
Contractor Follow-Up	Monitor	** ** ** ** ** ** ** ** **
Develop Procedures	1	**
Plan Traing Users	1	*
Complete Conversion Plan	1	**
Install Hardware	1	**
Install Operating System	1	**
Install Application Software	1	*
Train Personnel	1	** ** ** ** ** ** ** ** **
Project Status	1	* **
Develop Conversion Procedures	1	**
Purify Conversion Files	1	** ** ** **
Project Status	-	
User Proect Status	-	* *
Perform Integration Test	1	** ** ** ** ** ** ** *
User Validation	1	** **
Total Man Days	1	1 1 1 1 1 1 1 2 1 1 1 1 1 2 2 2 2 2 2

Chapter II

Information Center and Response Line

INFORMATION CENTER

An Information Center is an effective organizational tool for office automation organizations. There are many functions of such a group that can enhance productivity in all micro areas. These include the evaluation of new products, the assembly of useful application, the operation of an Information Center facility, cooperative consulting on user terms, a technical support group and several administrative activities.

In a large computer system installation, considerable advantage can be gained by establishing a cooperative central support group or Information Center. More analysis can be put into system development, more competent technical support personnel can be employed, and standardization can lead to advantages for all groups. An Information Center group should consider its prime objective to be the increase of productivity of the microcomputer groups. It is a service organization which applies rules, standards and procedures only to enhance the benefits for all within the company.

To be effective, the Information Center must be completely dedicated to all computer operations. The level of support required and the type of thinking required are different from technical support operations.

There are sufficient challenges, and a large field is opening up in the future. The technology is advanced and the possibilities are numerous.

The Information Center group could report to any of the major departments in an Information Services organization, or it could have a department of its own if it is of sufficient size. In practice, there will be numerous interactions with administration, systems, computer operations and communications, so the exact location is immaterial.

An Information Center group may or may not have an Information Center room to display its wares. If the organization is large and divided, a central place to exhibit equipment and software, to give advice and to give training, may be helpful and have the advantage of high visibility. On the other hand, if the organization is monolithic and has highly structured standards and procedures, there will be fewer selections that can be made by the users, and less purpose for an Information Center.

The Information Center concept has been found invaluable in servicing numerous user groups, in standardizing a limited number of computers for greater overall productivity and for useful organizational standards and procedures. It is an attractive and cost-effective approach to introducing computer systems and has proven to be popular and productive where tried. It is well worth consideration by any group if only to get volume discount prices on equipment that is acceptable to all.

The most important tool that the Information Center may have is what is termed the Response Line. Through this medium, the users do not feel isolated after the initial training. The Information Center becomes a hand-holding facility to the novice users until they become surer of themselves on the use of the system and software packages.

Some companies provide a WATS line for support of their products. It has been proven that this does not promote self reliance. The user knowing that the answer is just a free phone call away will almost assuredly rely on the Information Center for seemingly miniscule problems. This is not to say take away the user's help, but stress should be put on the importance of trying all avenues of a possible solution and as the last frustrating resort turn to the Information Center.

Response Line Support

Executive Briefing Overview

What is Response Line Support?

- On-Going Reinforcement of
    —Development
    —Documentaion
    —Training
    —Usability
    —Installation

- Tool for Building
    —Self-Sufficiency
    —Mental Ownership

- Done Remotely by Phone

- Demonstration of Corporate Commitment

Survey Results

—15 Calls/Users/Day
—12 Minutes
—50% Phone in Ear Time

X 2 Usability
X 3 Documentation

IMPACT ON RESPONSE LINE

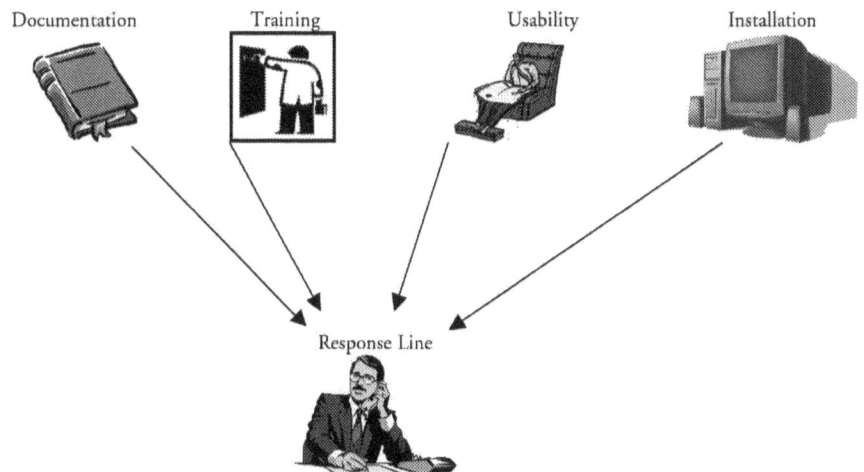

DEVELOPMENT

Documentation   Training   Usability   Installation

Response Line

Response Line Statistics
(Industry—Insurance Agency System)

- Currently: 800 Customers
                     37 Support People

- Calls By Type: Hardware   224
                     System Software   53
                     Application Software   858
                     Communications   418
                     Billings   20
                     Documentation/Training   646
                     Supplies   37
                     Installation   36

- Projections: 3800 System in 24 Months
                     176 Support People

- Results: 4400 Systems Installed
                     22 Support People

No.	Date	Time	WS / Name	Description of Problem	Branch Contact	Resolution	Date
1							
2							
3							
4							
5							
6							
7							
8							
9							
10							
11							
12							
13							
14							
15							
16							
17							
18							

Chapter III

Training

TRAINING

Development Steps

1. Confirm the initial user training.
2. Determine training requirements.
3. Determine training approach, for each session.
4. Define elements of training.
5. Determine standards for the training material
6. Organize the training material development activity.
7. Determine resources needed and/or available.

TRAINING APPROACH

The computer operations manager may help develop the training material for the computer operation personnel. Specialized training materials are sometimes available from the hardware, systems software and training vendors. Material should be evaluated to determine if it can be used directly or if it needs customizing.

Training session conducted by vendors can be directly applicable to a particular system. For effective user training, it may be necessary to operate the system in an environment that does not affect the live data base.

Classroom exercises provide familiarity with input forms and proficiency in performing repetitive operations.

Independent study may be appropriate for decentralized system users or user groups that are only slightly affected by the new system. These independent study units can be self-taught or with minimal instructor assistance.

The training material must directly reference the completed user documentation throughout each session.

In developing the training program the following should be documented:
1. Objectives of the training session.
2. Audience
3. Estimated length of each topic
4. The number and types of training sessions

TRAINING

The sessions may be organized by job or subject or functional classifications.

Seminars and small group meetings promote extensive two way communication. These are particularly effective for training management personnel.

Classroom instruction is more appropriate for training non-management users in the daily use of the system with discussion periods to explain specific points. It is during these periods that users should become familiar with documentation and reference aids.

Training must develop necessary skill and that users have a positive business attitude toward the system, not looking upon the system as drudgery, but on relieving drudgery.

User personnel should be trained to provide input and to use the output from the system.

Training should include both manual and terminal procedures. Inclusive of this training should also be a complete explanation of the system and the relationships of all input, output and error conditions. In this way personnel will be more self sufficient and be able to localize problems and take proper steps to correct.

Computer operations personnel should understand the operating environment and the processing requirements of the new system. Operations should also be familiar with utility routines, job control language and operator messages.

DEVELOPMENT OF TRAINING MATERIAL

Material developed in this task are the instructor guides, the training systems that run on the computer, audio visual materials and a full training sesssion agenda.

Using the system documentation **is** more economical that developing duplicate materials and minimizes in consistencies.

Instructor guides should include a description of the skill levels of the participants. An overall outline of the topics to be covered, including time allocations and copies of the user manuals.

Informative visual aids enhance the level of understanding and the effectiveness of the presentation.

The training materials should be structured with key consideration on the content. Each training session should begin with a clearly stated list of objectives that explains the roles of the participants.

Pilot training is done by an instructor with actual developed materials and representative participants. The environment should be identical to that used to conduct live sessions.

Train the trainer sessions may be necessary. These would teach presentation techniques and familiarize instructors with the subject matter.

At the end of a completed session a handout of the session or course evaluation should be provided. This will show areas of weakness or strengths. The evaluation should encompass; the instructors, materials, exercises and appropriate achievement of course objectives. There should also be a section to include overall comments from participants.

Trading Objectives

- Users learn as Quickly as possible
- Users are able to perform on the job
- Available when needed
- Promotes self-sufficiency
- Cost effective

Classroom Instruction

- Must be structured
- Provide frequent breaks
- Vary the media Interject humor
- Consider Length of day
- Provide meaningful exercises
    —Have more than required
- Allow time for question
- Provide handouts—Reference book

Self-Study Materials

- "How-To" Manual
  - —Leads the user step-by-step through a task
  - —Hands-on practice

- Reference material
  - —Supplemental to "How-To" manual
  - —Provides detail

- Training Administrator guide
  - —Explains material, intended audience, etc.
  - —Provides roadmap for training

Prioritize Locations

- Size
- Business risk
- Geographic proximity
- Unique functions
  —Personality
  —Talent
  —Experience
  —Desire to change

Configuration
  Current system
  Politics
User scheduling considerations
  —month-end closing
  —physical inventory
  —vacations
  —major systems or application changes

Inhibitors to Learning

- Pressure on Student

- Unclear direction

- Motivation

- Learning Pace

- Instructor Burnout

- Too Much Help

- Too Much Information

- Inappropriate Exercises

Chapter IV

Documentation

DOCUMENTATION PLAN

Developmental Steps

1. Gather the necessary documentation for each manual
2. Identify any additional procedures required.
3. Package the documentation, ensuring consistency.
4. Determine how the manuals are to be produced.
5. final forms design.
6. Finalize delivery dates.
7. Negotiate orders with the supplier of the forms.
8. Identify additional supplies required.
9. Procure supplies.
10. Develop periodic reordering procedures.

If the procedures for the new system are written as updates to existing manuals, the content, style and organization must be consistent with the existing manuals.

Type of Manuals

1. Management Overview
2. Terminal Operations
3. Security and Control Information
4. Contingency Planning or Fallback Procedures
5. Help Desk Manuals
6. Production Control Procedures
7. Computer Operations Manuals

Documentation = Information

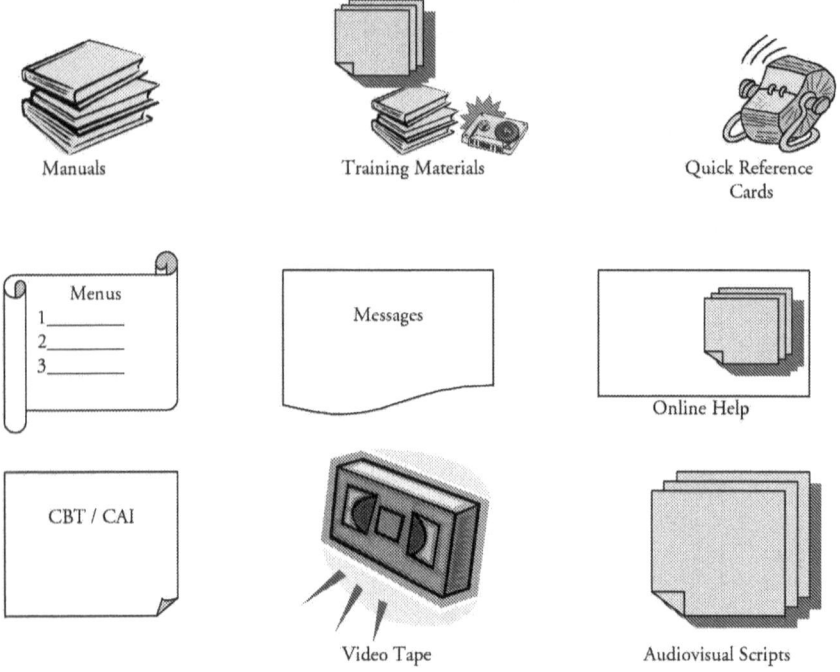

Manuals

Training Materials

Quick Reference
Cards

Menus
1_____
2_____
3_____

Messages

Online Help

CBT / CAI

Video Tape

Audiovisual Scripts

SUPPORT DOCUMENTATION

The next pages are an example of an imaginary documentation. This will give the reader a general under standing of what a finished in-house documentation package should resemble. This particular example is derived from the methodology that IBM follows in producing their equipment instruction documentation.

The structure of the manuals will vary according to the intended user and the scope of the system.

Step 5—Enter Buyer/Seller Informarion

Screen Description

The Buyer/Seller Detail screen is accessed by pressing CMD5 from the Order Entry Header screen, the Order Entry Lender/Realtor screen, the Order Entry Property screen, or the Order Entry Comments screen.

The Buyer/Seller Detail screen is used to input information about buyer(s), seller(s), and owner(s) of record of the property in the question.

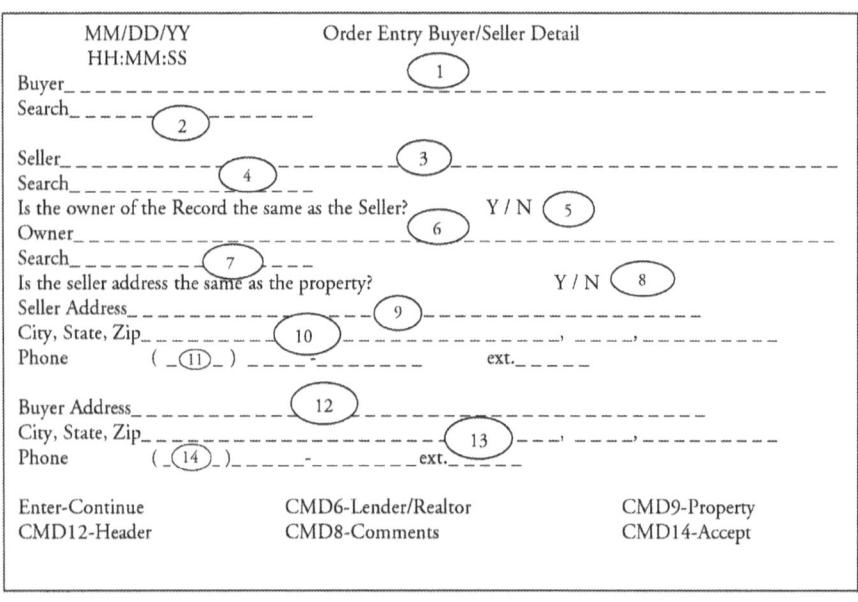

Fig. 402-5

Field Description

1. Buyer—Type the name of the buyer.

2. Search—???

3. Seller—Type the name of the seller.

4. Search—???

5. Owner of the Record Same—Type Y if the owner of the record is the same as seller; type N if the owner of the records is the other than seller.

6. Owner—Type name of owner if the record of record is other than seller (see field 5).

7. Search—???

8. Seller Address Same?—Type Y if seller's address is the same as the property address; type N if seller's address is the other than the property address.

9. Seller Address—Type seller's city, state and zip.

10. City, State, Zip—Type seller's city, state, and zip.

11. Phone—Type seller's phone number.

12. Buyer address—Type buyer's address.

13. City, State, Zip—Type buyer's city, state, and zip.

14. Phone—Type buyer's phone number.

If You Want To	Then	Result	Next Step
Enter Buyer/Seller Information	Place the Cursor at the beginning of the field you wish to enter by pressing TAB and typing the information		Continue to next field by pressing TAB
Enter Property information	Press CDM9, TAB to each field to continue	Order Entery property Detail screen appears	Go to step 4
Enter Lender/Relator Information	Press CDM6, TAB to each field to continue and press ENTER. Press CMD12 to return to the header screen	Order Entry Lendor/Realtor Detail screen appears	Go to step 6
Enter Order Comments	Press CMD8 and enter order comments	Order Entry Comments Detail screen appears	Go to step 7
Return to Order Entry Header screen	Press CDM12	Order Entry Header screen appears	

Chapter V

Usability

PROJECT USABILITY

Conduct a usability evaluation to evaluate:

—Ease of use of Software Application
—Effectiveness of Training
—Ease of Documentation

A Usability Evaluation is a method of evaluating how all the elements of the system will work in a typical end-user environment.

Risks of Not Evaluation Usability

— Loss of Credibility

— End user Dissatisfaction

— Fewer Systems Installed

Benefits of Usability

— Produce Higher Quality Product

— Increase End-User Acceptance

— Reduce Revision and Support Costs

— Improve Effectiveness of Product Year

— Apply New Techniques/Ideas to Future Products

Chapter VI

Security and Disaster Recovery Preparation

SECURITY AND DISASTER RECOVERY PREPARATION

A critical part of the administration of the Information Center is focused on the preparation for security and privacy controls, and the planning for the recovery from a disaster. These aspects are frequently considered of secondary importance in the eagerness to get the Information Center started and to offer as much as possible to the users. If breaches of security or privacy reach management's attention, however, or if even a minor disaster strikes, such as a local fire, great damage can be done to the perception of the Information Center and to its ability to perform its function adequately.

The objectives of security standards and resources protection preparations are to make sufficient preparations to hold safe the organizations' resources that are involved with computing activities, and to make all end users and staff aware of the necessity for measures for security and resource protection.

There are always several levels and types of security that are necessary. Some measures have to do with physical protection of resources and other measures have to do with the operational and audit controls that are established in the software. It should be clear that the resource protection required always includes the organizational data that is in the possession of end users, and any vital information derived from it.

Physical security includes equipment, methods and plans to protect the personal computer system, peripheral equipment, storage media, system documentation, and applications software.

One of the principal problems in office automation is being assured of the accuracy of the information that is made available. Seldom does the computer make errors, but frequently the users of information will change the data to suit their own analysis or reporting needs, and then mix it with controlled data. Businesses strive to maintain the accuracy of the computerized data, but it is very difficult once it has left the mainframe. Data may be entered or altered incorrectly without sufficient controls. The need to apply substantial controls will depend completely upon the value and need for the integrity of the information.

Privacy must also be considered as data is communicated and shared. Privacy refers to the rights of individuals and organizations to determine for themselves when, how, and to what extent information about them is to be made available to others. In some cases, there are laws requiring the necessary degree of privacy to be managed, while in other cases, it is simply up to good judgment.

The requirements for managing security in the computer environment will be specific to each instance and will need to be analyzed by the Information Center staff or another appropriate group. It is important to remember that any security hardware or software options, or physical site security, should be cost justified. Many smaller operations cannot afford the cost of advanced EDP security devices, or security audit procedures that are available to larger organizations. It would just not be cost effective to install $30,000 worth of physical security to protect a $40,000 computer configuration that does not contain highly sensitive, valuable data.

The common sense approach to security needs is the best approach. Adequate security can be achieved through the balanced employment of security administration, physical and software security features, risk analysis, disaster planning, and awareness training of staff and end users.

Disaster recovery planning arrangements for contingency backup operation are important. A useful approach is to relate the cost of different levels divided of security and disaster recovery measures can be divided into mandatory and desirable categories, and appropriate procedures can be applied to each. The cost should be analyzed, and the choice presented to management.

The purpose of disaster recovery planning is to prepare in advance to ensure the continuous of business information if any EDP capability is lost. Disaster recovery preparation is a management issue, rather than a technical issue. They deal with the realities of people, organizational relationships, and special interests. Disaster recovery actions are highly prioritized and many normal operations may be neglected. Management must take the lead and continually assess the technical considerations involved, as to the utility of preparedness measures.

Chapter VII

Installation

Installation/Support Plan

Elements   Responsible _____

Hardware
  —Warranty and Maintenance   _____
  —Different Configurations   _____
  —Different Vendors   _____
  —"Burn In" Testing   _____
  —Upgrade Procedures   _____
  —Engineering Changes   _____
  —Central Maintenance vs. Decentralized   _____
  —Who "Installs" Equipment   _____
  —Checklists   _____

Site Preparation (Physical Planning)
  —Safety   _____
  —Electrical Considerations   _____
  —Heating/Air Conditioning   _____
  —Furniture   _____
  —Cabinets, Diskette Storage   _____
  —Telephone   _____
  —Checklists   _____

Conversion
  —Manual to Automated   _____
  —Training   _____
  —Parallel Operation   _____
  —Checklists   _____

Supplies
  —Initial...How Much...Who Pays For   _____
  —On-Going...Lead Time to Receive   _____
  —Supplies Catalog   _____
  —Checklists   _____

Chapter VIII

Project Manager

Planning Schedule

Project:		
Name:	Date Due:	Page:	Of:

Activity	F	A
	F	A
	F	A
	F	A
	F	A
	F	A
	F	A
	F	A
	F	A

F = Forecast
A = Actual

Installation Management
<u>Keys To Success...</u>

° End user self-sufficiency

° Project support Plan

° Pilot Installation

Installation Management
The Plan

° Should address outstanding issues...
responsibility...dates

° Should document agreed-to plans...
responsibility...dates

° Must contain contingency plans

Installation Management
<u>Mission</u>

° Attain your project goals

° In shortest time

° Using available resources

° To achieve end user satisfaction

Installation Management
The Pilot

Purpose:

- ° Initially validate elements and timings of the plan

- ° Define need for additional planning or pilot tests

Installation Management
The Pilot

Elements:

- Define completion criteria

- Conducted as near to reality as possible

- Monitor results...but don't assist!

- Should validate important elements of the plan
    —adequacy of training
    —thoroughness of documentation
    —user support (IBM, vendors, corporate)
    —physical planning/site readiness model

- Lab environment or user site

Installation/Support Plan
Elements & Responsibility

- ° Hardware
- ° Sit Preparation
- ° Conversion
- ° Supplies
- ° Software
- ° Change Control
- ° Documentation
- ° Pilot Testing
- ° Response Line
    —help desk
    —hotline

Chapter IX

Risk Knowledge

Control System

	Diversity of Data Use	Accessibility	Sharing of Business Systems	Data Independence	Competency of People	Selection of Computer Technology	Systems Development Process	EDP Controls
EDP Risk[2]								
Improper use of Technology	1	2	2	3	2	3	3	1
Inability to react quickly	2	1	2	3	2	3	3	2
Inability to control technology	2	3	2	3	3	2	2	2
Differences[3]								
Human functions replaced with machines					v	v	v	v
Coded data not readable by people					v			v
Rapid processing			v	v	v			v
Error preprogrammed	v			v	v		v	v
Automation of Control		v	v	v		v	v	v
Centralization of functions	v		v	v		v	v	v
New forms of evidence		v		v		v		v
New Methods of authorization		v		v	v	v		
New proccesing concepts				v		v	v	

[2] 3 = High need for control, 2 = average for control, 1 = low need for control

[3] v = difference affects control

	SDLC					
	Feasibility	System Design	Program	Test	Operate	Feedback
EDP Risk[2]						
Improper use of Technology	1	3	2	1	1	3
Inability to react quickly	1	3	1	1	3	3
Inability to translate needs into technical requirements	3	3	2	1	1	2
Inability to control technology	1	3	3	3	3	2
Differences*						
Human functions replaced with machines	v	v				v
Coded data not readable by people		v				v
Rapid processing		v	v	v	v	v
Error preprogrammed		v	v	v	v	v
Automation of Control		v	v	v	v	v
Centralization of functions	v	v				
New forms of evidence	v	v				v
New Methods of authorization	v	v				v
New processing concepts	v	v	v	v	v	v

[2] 3 = High need for control, 2 = average for control, 1 = low need for control

[3] v = difference affects control

Bibliography

Alpor, Alan. "Ford Setting Up $300 M. on Pact."
    Computerworld, 23 June 1986, pp. 1.

Angus, Ian. "The Myth of the Supercontroller."
    Computer Decisions, 28 October 1986, pp. 34.

Bacas, Harry. "Sharing the Load."
    Nation's Business, October 1985, pp. 56.

Baker & McKenzies Information System. "Office
    Automation Through Word-Processing Terminal."
    The Magazine of Office Administration and
    Automation, April 1987, pp. 11.

Beaver, Jennifer. "Ford Misgives OA Keys to IBM."
    Computer Decisions, 18 September 1986, pp. 18.

Beaver, Jennifer E. "The 937X and OA: Time Will
    Tell All." Computer Decisions, 4 November 1986,
    pp. 17.

Beeler, Jeffry. "HP Adds to Office Information System
    Family." Computerworld, 8 December 1986, pp. 13.

Bridges, Linda. "PC Pilot Programs Test Automation
    Approaches." Incorporate, 1986, pp. 40.

Call, Barbara. "In-House PC Expose."
    PC Week, 19 August 1986, pp. 51.

Canning, Donnia. "25 Guaranteed Ways to Fall."
    Administrative Management, December 1986, pp. 14.

Cantrell, Wesley and Johnson, Donald R. "The Office
    of the Future." American School & University,
    September 1985, pp. 24.

Coggshell, William L. "Integrated Software New Power
    for OA." Modern Office Technology, 30 April 1985,
    pp. 51.

Cox, James F. "Education, Training needs must be
    assessed before system implementation", Data Management,
    24 May 1986, pp. 10.

Czubeck, Donald H. "Understanding IBM's Electronic Mail
    Architecture." Data Communications, 15 November
    1986, pp. 157.

Derrons, Donna E. "The Paperless Office."
    Information Evolution, 10 November 1986, pp. 56.

Diebold, John, "Remembering Automation."
    Computerworld, 3 November 1986, pp. S126.

Dowst, Somerby. "Where is the Office Automation Going
    Anyway." Purchasing, 20 February 1987, pp. 40.

Dun's Business Month. "Office Automation Advances White
    Collar Productivity." Dun's Business Month Focus,
    March 1986, pp. 59.

Dykeman, John B. "Requiem for the White Collar
    Workforce." Modern Office Technology,
    31 July 1986, pp. 12.

Dyson, Esther. "The Factory in Your Office." Forbes,
    25 January 1988, pp. 94.

Editorial Staff of University of Chicago Press.
    "The Chicago Manual of Style, 13th Edition."
    University of Chicago, May 1986.

Evans, Sherli. "Workstations that Complement Computers."
    Office Administration and Automation, May 1985,
    pp. 42.

Faust, Maureen. "Software Gets Stationary Desktop Accessory."
    PC Week, 3 November 1986, pp. 163.

Fleming, Marilyn. "Software for the Automated Office."
    The Office, September 1987, pp. 40.

Fellmy, William R. "Make Your Peripherals Perform."
    American School and University, October 1987, pp. 18.

Godwin, Nadine. "Study Finds 32% of Retailers Own Back
    Office Automation Systems." Travel Weekly,
    September 1986, pp. 1.

Goldstein, Mitchell. "Managing Office Automation."
    National Productivity Review, Spring 1985, pp. 184.

Goldstein, Mitchell. "Managing Office Automation Part VI."
    National Productivity Review, Summer 1985, pp. 250.

Hamilton, Clark. "Neither Paperless nor Personless."
    Association Management, October 1987, pp. 47.

Hamilton, John W. "Measuring Technology."
    Journal of Systems Management, November 1987,
    pp. 19.

Hansgate, John P. "Office Automation: The User Profile."
    The Office, August 1985, pp. 19.

Heckel, Paul. "The Elements of Friendly Software
    Design." Wadsworth Publishing Co., 1984.

Hess, R.D. and D.F. Walker. "Instructional Software,
    Principals & Perspectives for Design & Use."
    Wadsworth Publishing Co., 1985.

Hohner, Gregory. "Ten Issues to be Resolved in Office
    Automation." Information Management, April 1985,
    pp. 9.

Johnson, Don S. "Gazing Through a Dusty Crystal Ball:
    What's Ahead in OA." Administration Management,
    January 1986, pp. 7.

Johnson, Don S. "The Word for Today is Innovative."
    Administrative Management, June 1986, pp. 5.

Johnson, Don S. "The Immeasurable Benefits of Office
    Automation." Administration Management,
    July 1986, pp. 5.

Johnson, Don. S. "The Systems Approach to Office
    Operations." The Magazine of Office Administration
    and Automation, March 1986, pp. 5.

Kalb, Bill. "Factory Automation Computer Technology
    Systems." Automotive Industries, January 1987,
    pp. 68.

Kalb, Bill and Zoia, Dave. "Ford Nears Pact on
    Computer Net." Automotive News, 30 June 1986,
    pp. 1.

Kleinschrod, Walter A. "A Conversation with John
    Connell." Office Administration and Automation,
    June 1985, pp. 30.

Kropper, Jon F. "Office Automation and the
     Factory-of-the-Future." Design News,
     April 1985, pp. 15.

Lacob, Miriam. "Survivors of 1990."
     Computer Decisions, 15 January 1985, pp. 74.

Lanzillotti, Victor F. "Installation: It's Not
     What You Think." Bottomline, 3 February 1986,
     pp. 35.

Laszlo, George. "Ten Steps to Successful Systems Design."
     The Magazine of Office Administration and
     Automation, October 1986, pp. 23.

Luckert, Laura and Maddox, Addie. "All the Right Moves;
     How to Win When Your Company Goes On-Line."
     Women in Business, July-August 1986, pp. 16.

Ludlum, David. "Faith Required to Invest in
     Productivity." Computerworld, 3 November 1986,
     pp. 195.

Mackie, James. "Office Automation: Putting Integration
     Into Perspective." Telecommunications, 19 May 1985,
     pp. 44.

Masud, S.A. "GAO Urges; Suspend Funding on Planning
     Res. - Patent Pact." Electronic News, 4 August
     1985, pp. 43.

Morgan, James P. "Business Transactions and How They
     Are Handled by the System." Purchasing,
     26 February 1987, pp. 77.

Murljacic, Tony. "The Evolution of Automated Office
     Systems." Infosystems, August 1987, pp. 40.

Niles, John S. "Automated Geographic Data Enhances
     Many City Functions." Nation's Cities Weekly,
     14 September 1987, pp. 11.

O'Connor, Bridget. "Learn How to Learn."
     Administrative Management, May 1987, pp. 13.

Ostrowski, John W. "Microcomputer Management Support
     Strategies." The Bureaucrat, Summer 1986,
     pp. 53.

Papper, Jon. "Homemade Software." PC Week,
    12 August 1986, pp. 45.

Paznik, Jill. "Making People Feel Comfortable with
    Technology is a Manager's Job."
    Administrative Management, August 1986, pp. 8.

Porter, Robert W. and Trill, Gilbert T. "A Little
    Computer Can Do a Lot of Work!" Purchasing,
    23 May 1985, pp. 83.

Pyykkonen, Martin. "Plan for Your Lan." Datamation,
    15 October 1987, pp. 109.

Rosen, Arnold. "Administrative Management." The
    Magazine of Office Administration and
    Automation, June 1987, pp. 6.

Ruby, Daniel. "Methods to Measure PC's Usage, Benefits."
    PC Week1, 10 March 1986, pp. 40.

Sager, Ira. "IBM S-36 Falls Short of Goal."
    Electronic News, 6 January 1987, pp. 1.

Sandburg, Dorothy. "Trends in Technology."
    Administrative Management, January 1986, pp. 10.

Seither, Mike. "Office Software Spans Multiple
    Operating System." Mini Micro Systems,
    April 1986, pp. 47.

Stamps, David. "Pioneering: It Hasn't Been Easy."
    Datamation, 15 February 1986, pp. 62.

Steinberg, Don. "Anatomy of a Connectivity Contract."
    PC Week, 24 February 1987, pp. 3.

Straussmann, Paul A. "Office of the Future."
    American Printer, January 1985, pp. 53.

Sullivan, Michael. "Integrated OA Systems."
    Computerworld, 30 June 1986, pp. 33.

Sutherland, Duncan. "The Future of the Office."
    Management Review, July 1986, pp. 44.

Tapscott, Don. "OA Banks on Connectivity."
    Datamation, 15 March 1986, pp. 106.

Tucker, Peter. "Right Tool for the Right Job."
    Products Spotlight, February 1988, pp. 52.

Verity, John. "Grafting a Wonder Office System."
    Datamation, 15 March 1986, pp. 88.

Wardell, David. "Office Management System Offers
    New Techniques." Travel Weekly, 22 September
    1986, pp. 106.

Weiszmann, Carol. "Office Information Systems."
    Industry Week, 27 May 1985, pp. 45.

Yourdon, Edward. "Paper Chase." Computerworld,
    21 July 1986, pp. 53.

0-595-30690-X

www.ingramcontent.com/pod-product-compliance
Lightning Source LLC
Chambersburg PA
CBHW030915180526
45163CB00004B/1839